SYMBIOSIS | ˌsɪmbɪˈəʊsɪs, -bʌɪ-|

noun (pl. symbioses |-siːz|)

    interaction between two or more different organisms living in
    close physical association, typically to the advantage of each.

Over the past 4 billion years, microbes have
shaped our Earth into the biosphere we know and
love – rich in biological and geological diversity.

Through a range of symbioses (some brief, some
lifelong), microbes have collaborated with all types
of life on Earth to create new, emergent forms,
including human beings. While some symbioses
cause harm, most bring benefits to all involved.

The idea that life evolves through competition is
only part of the story. Life is just as much about
working together.

Created by Briony Barr & Dr. Gregory Crocetti
Illustrated by Aviva Reed
Written by Ailsa Wild
Inspired by the research of Prof. Madeleine van Oppen
Art Direction, Scientific Illustration & Storyboarding by Briony Barr
Scientific Writing by Dr. Gregory Crocetti & Beth Askham
Researched by Dr. Gregory Crocetti & Prof. Linda Blackall
Book Design by Samantha McFadden & Jaye Carcary

Thank-you to all the people who supported our original Pozible campaign.
Special thanks to Kate Green, David Bourne and Janice Lough from
the Australian Institute of Marine Science, Gareth & Natalie Phillips at
Reef Teach, ArtPlay, Merry Youle, Jane Sullivan, Margaret Kett, Jill Farrar,
Veronica Radice, Sandhya Drew, Peter Barr, David Suzuki, George Aranda,
Cheryl Power, Jenny Martin, Theresa Harrison, Elaine Crocetti, Tom Danby,
the year 5 students at Thornbury Primary School (2014) & year 3 students
at Preston Primary School (2017) …and our mascot Kaisha Reed.

The creation of this book was supported by the Victorian Government
through Creative Victoria and the Australian Society for Microbiology.

This book is dedicated to the memory of the Australian poet
Judith Wright (1915–2000), who worked collaboratively with
artists, activists and ecologists to protect the Great Barrier Reef.

A catalogue record for this book is available from the National Library
of Australia.

Published in collaboration with Scale Free Network by

CSIRO Publishing
Locked Bag 10
Clayton South VIC 3169
Australia
Telephone: +61 3 9545 8400
Email: publishing.sales@csiro.au
Website: www.publish.csiro.au

Printed in China by Toppan Leefung Printing Limited

The views expressed in this publication are those of the author and do
not necessarily represent those of, and should not be attributed to, the
publisher or CSIRO.

How to pronounce the names of the characters in the story:

AMOEBA	ah-MEE-bah
COPEPOD	KOH-per-POD
CY	sy
CYANOBACTERIA	sy-A-noh-bak-TEE-ree-ah
DARIAN	DAH-ree-en
DINI	DI-nee
RHIZOBIA	ry-ZOH-bee-ah
ZOBI	ZOH-bee
ZOOX	zooks

PHOTO CREDITS:
Pg2   Earth & QLD Coastline: images courtesy of NASA
Pg34   QLD Coastline (zoom): image courtesy of NASA
        Coral Bommie: image courtesy of LTMP team, AIMS
        Coral Colony: image courtesy of LTMP team, AIMS
        Coral Polyp (close-up): image courtesy of Jean-Baptiste Raina, AIMS
Pg35   Polyp: Photo from the Okinawa Institute of Science and Technology
Pg37   Zooxanthellae: image courtesy of Cathy Liptrot, AIMS
Pg39   Zobi: Brennan et al/eLife 2014;3:e01579/CC BY 3.0
        Cyanobacteria: micrograph by Samuel Cirés
Pg40   Mucus Micrograph: Image courtesy of the Lewis Lab at
        Northeastern University. Image created by Anthony D'Onofrio,
        William H. Fowle, Eric J. Stewart and Kim Lewis

ZOBI AND THE ZOOX
A STORY OF CORAL BLEACHING

Written by Ailsa Wild
Illustrated by Aviva Reed
Created by Briony Barr & Dr. Gregory Crocetti

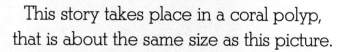

This story takes place in a coral polyp,
that is about the same size as this picture.

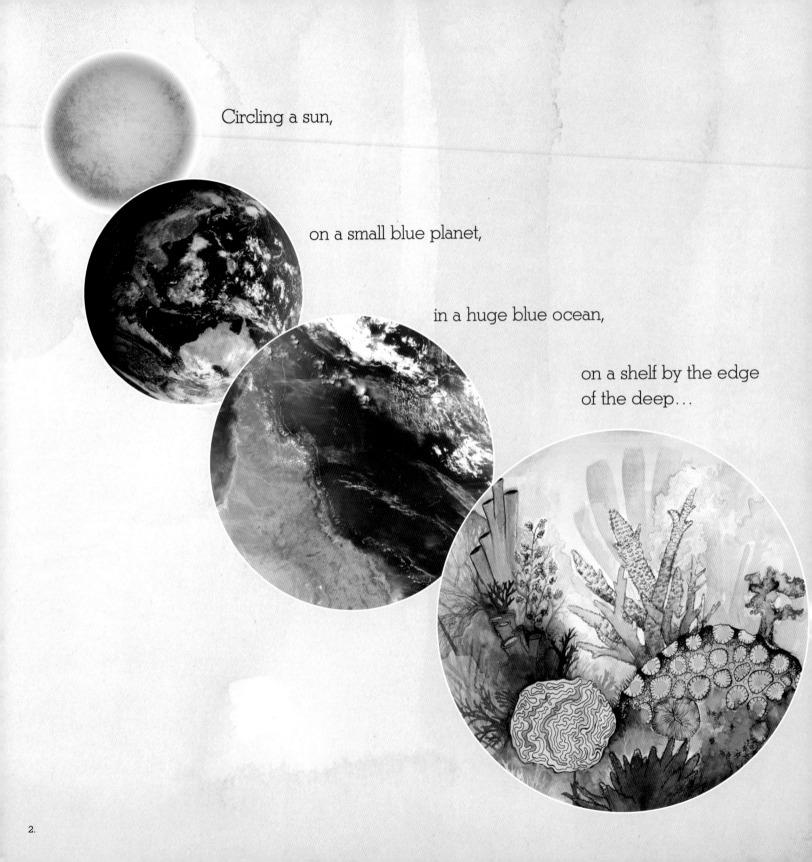

Circling a sun,

on a small blue planet,

in a huge blue ocean,

on a shelf by the edge
of the deep…

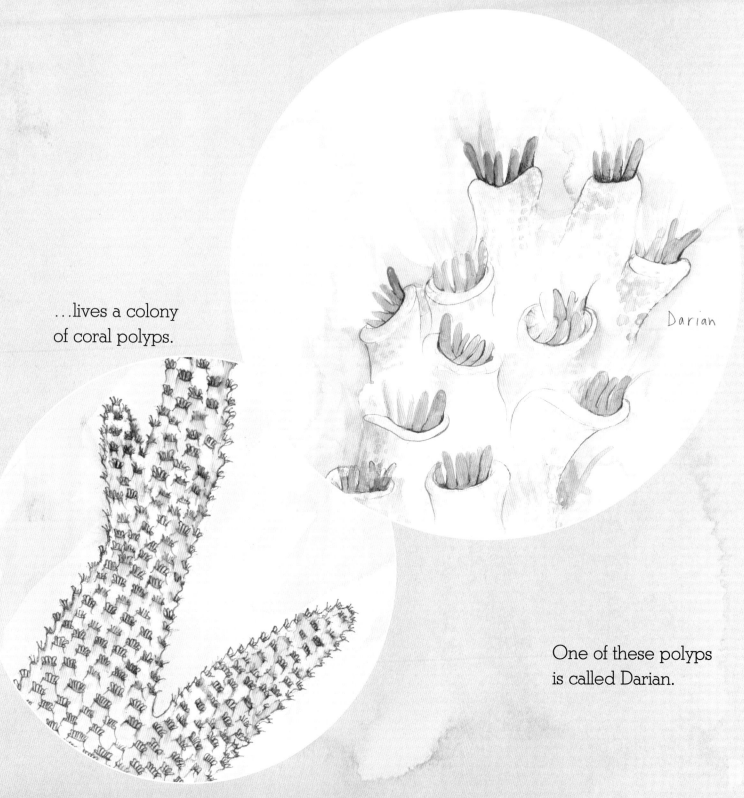

...lives a colony of coral polyps.

One of these polyps is called Darian.

Darian

Darian is building the reef, layer by layer,
like his family has done for hundreds of years.

10 years ago
20 years ago
30 years ago

4.

Trillions of small friends live in and on Darian. They usually work together in a happy, busy balance.

But not today.
Today, there's a problem.
The ocean is still and it's far too hot.

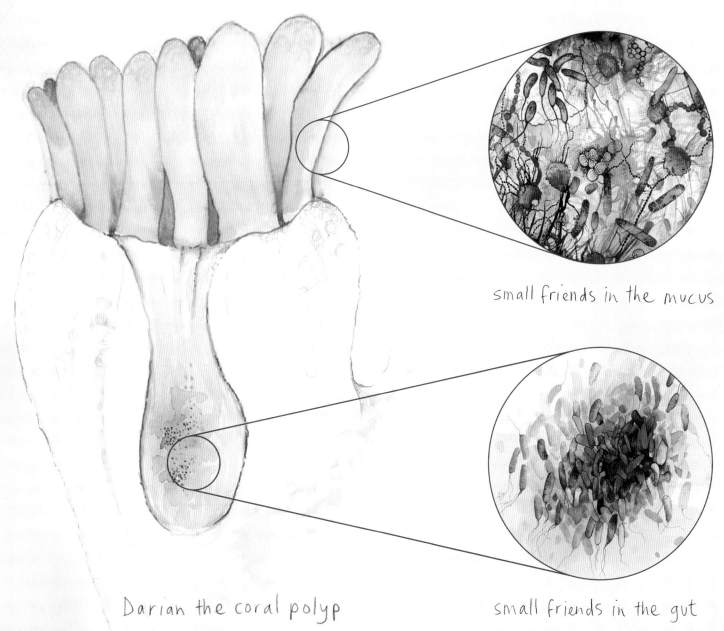

small friends in the mucus

Darian the coral polyp

small friends in the gut

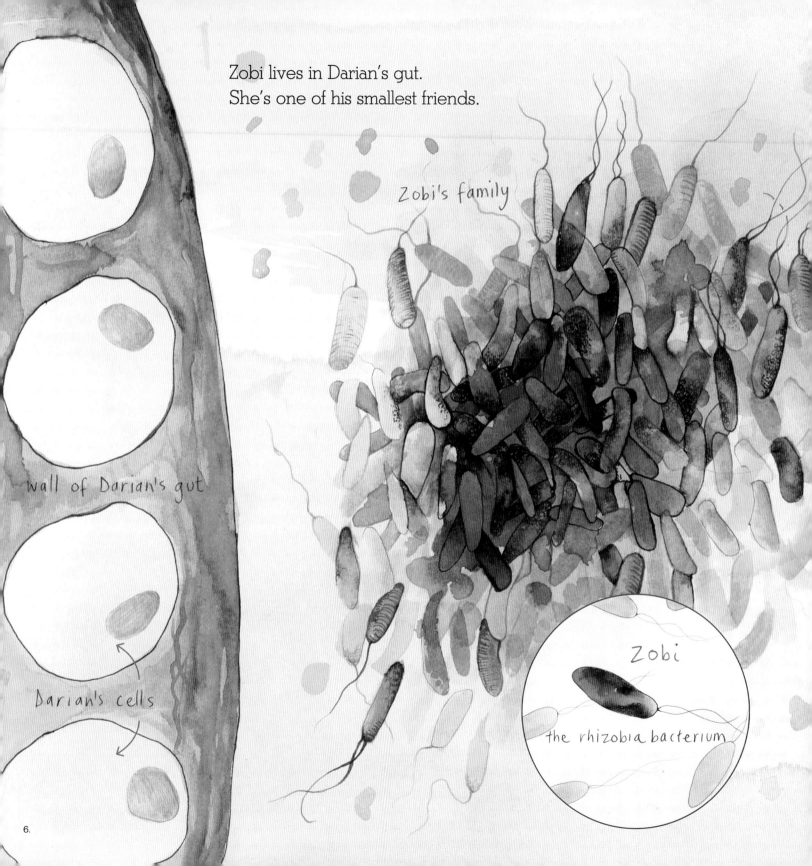

Zobi lives in Darian's gut.
She's one of his smallest friends.

Zobi's family

wall of Darian's gut

Darian's cells

Zobi

the rhizobia bacterium

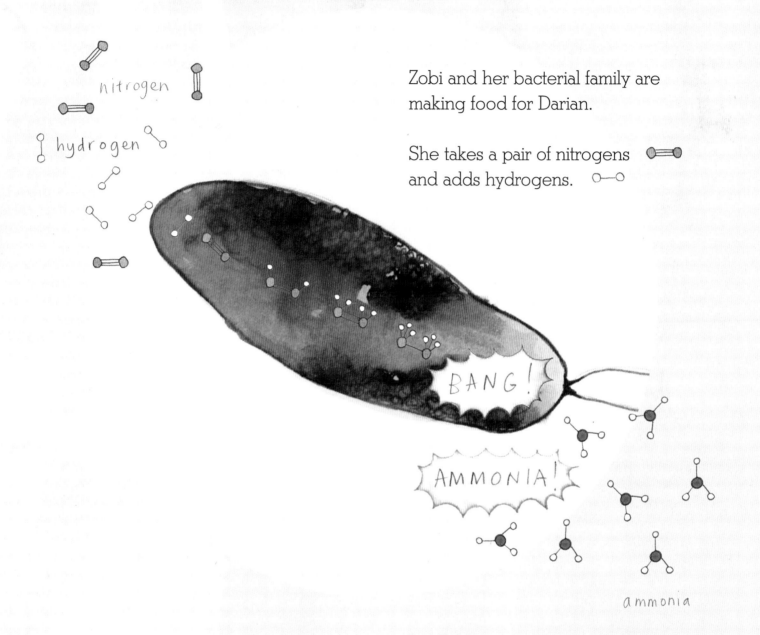

nitrogen

hydrogen

Zobi and her bacterial family are making food for Darian.

She takes a pair of nitrogens and adds hydrogens.

BANG!

AMMONIA!

ammonia

But Zobi can feel that something is wrong.
"Something horrible is happening," she says.
"Why is it so hot?"
Her family is too busy to reply.

So Zobi sneaks away…

Zobi treks up out of Darian's mouth, into the branching strands of mucus, which wrap around Darian like a snotty, protective blanket.

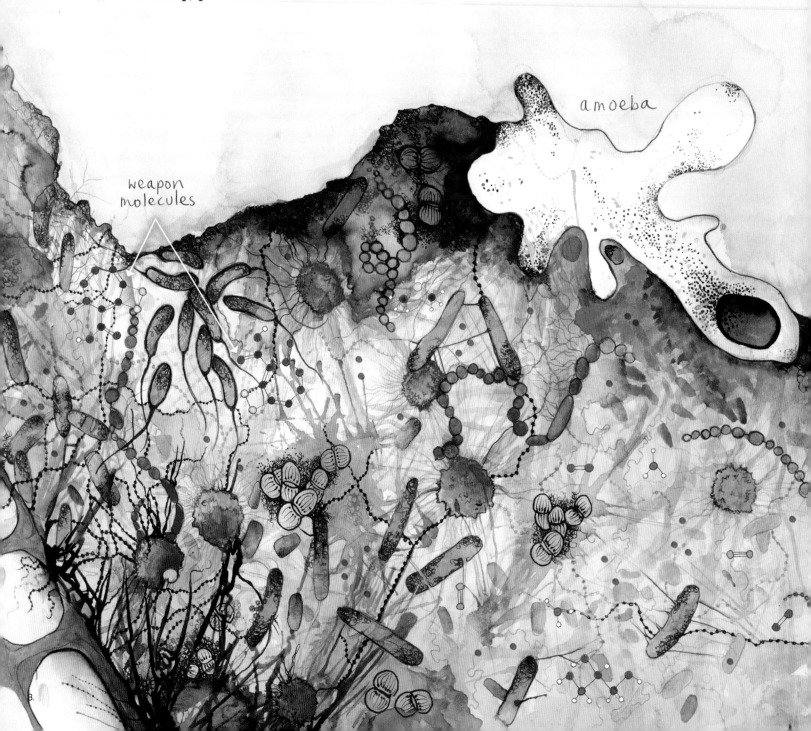

weapon molecules

amoeba

It's so busy in the mucus. Zobi dodges between microbes who are eating, recycling and swapping molecules. She passes a crew of guardian bacteria making deadly weapons to defend Darian.

She steers clear of a grazing amoeba, and wriggles up towards the sunlight.

From up here, she can sense the wild ocean, where anything might happen.

greeting molecules

edge of mucus

Zobi finds Cy in a sunny part of the mucus.
Cy is wise and old.
"What's going on?" Zobi asks. "It feels like
something terrible is about to happen."
"You're right, little Zobi," Cy says. "Trouble
does lie ahead."

This makes Zobi even more afraid.
"What?" she asks. "What trouble?"
"Have you heard the story of Old Pora?"
Cy begins.

Cy the cyanobacteria

Zobi

Zobi has heard the story. Just there, through the blue, is the skeleton of a coral colony. A broken pile, covered with thick, dirty algae.

Cy whispers, "Before Old Pora died, the water was hot for weeks. Just like today." Cy leans in to Zobi.
"We need to check the zoox," she says.

Old Pora

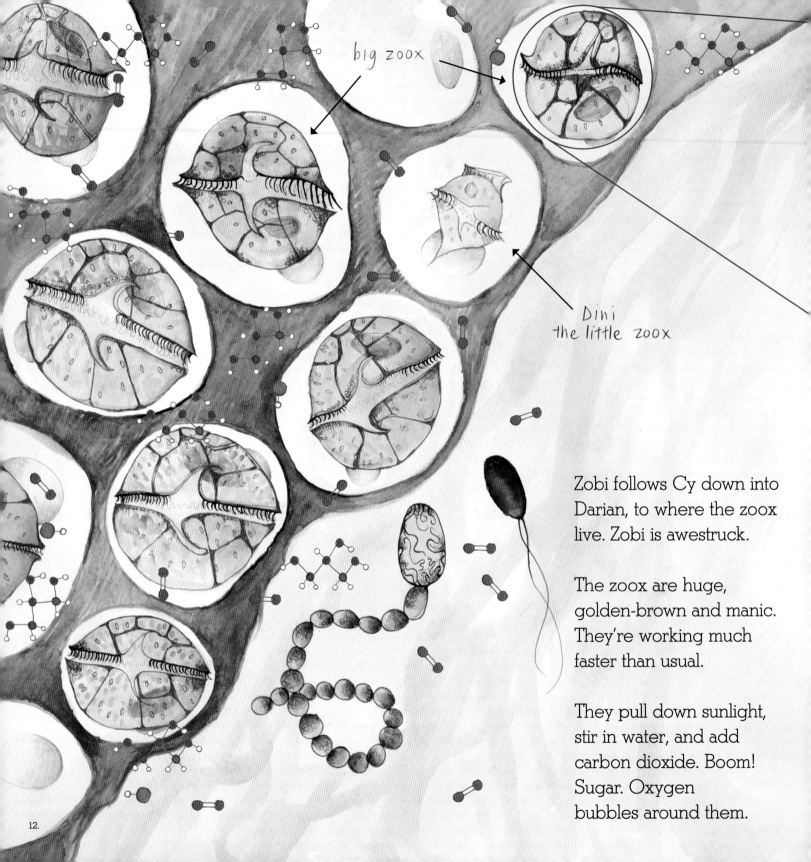

big zoox

Dini
the little zoox

Zobi follows Cy down into Darian, to where the zoox live. Zobi is awestruck.

The zoox are huge, golden-brown and manic. They're working much faster than usual.

They pull down sunlight, stir in water, and add carbon dioxide. Boom! Sugar. Oxygen bubbles around them.

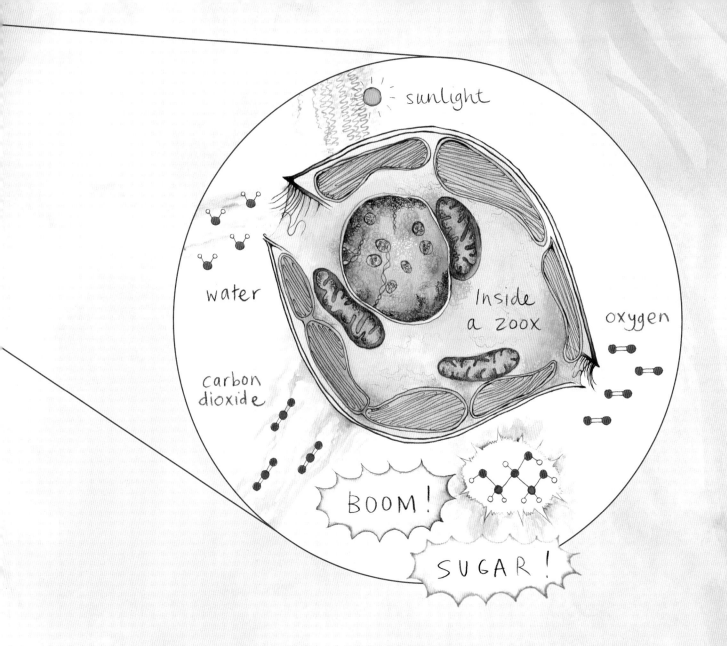

Zobi notices one zoox nearby, smaller than the rest. She's working slowly.

"That's Dini," says a big zoox, meanly.
"She's one of the Littles. They're always slow."
Dini carefully mixes water and carbon dioxide.

The busy zoox chorus their disapproval.
"Littles are slackers."
"Not like us. We work fast."
And the big zoox *are* working fast.

Carelessly fast.

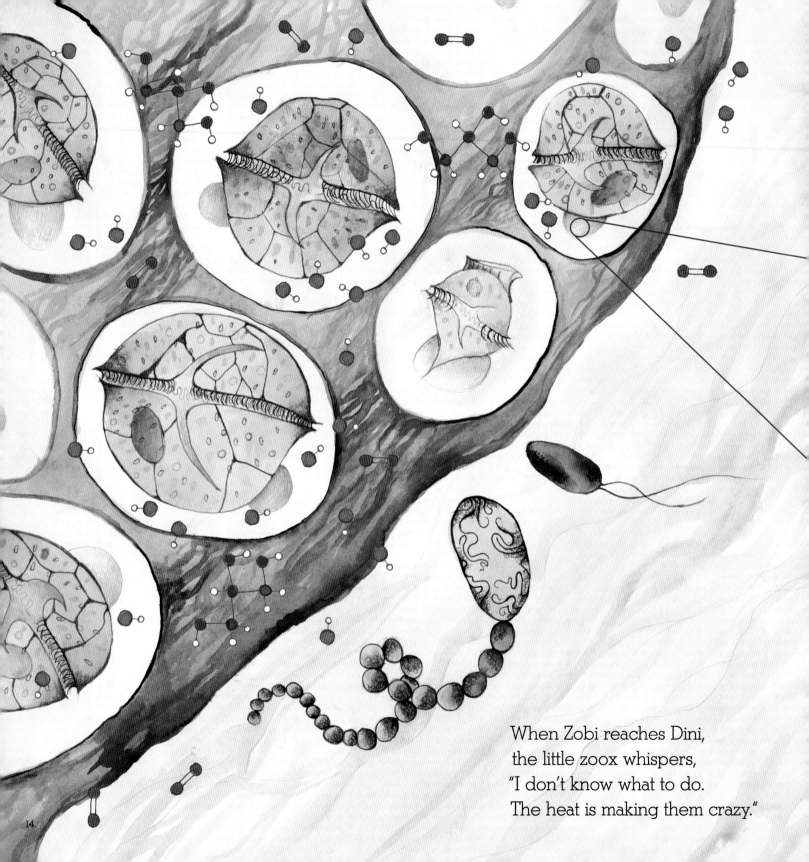

When Zobi reaches Dini,
the little zoox whispers,
"I don't know what to do.
The heat is making them crazy."

14.

A tiny toxic molecule zooms past Zobi and smashes up a protein. Wham! Another breaks a strand of DNA.

The nasty molecules are hurting Darian, and Zobi realises they're coming from the big zoox.

"One long, hot summer, Old Pora's zoox went crazy just like this," Cy says. "She had to eject them. Old Pora was a splendid golden-brown until she kicked out her zoox. It was a disaster. She didn't survive."

Zobi is afraid.
"We need to do something," she says.
But what can they do?
The heat is almost unbearable.

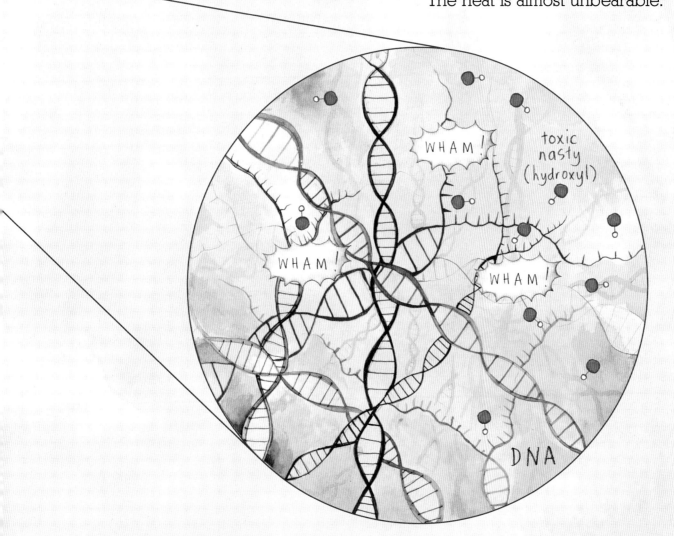

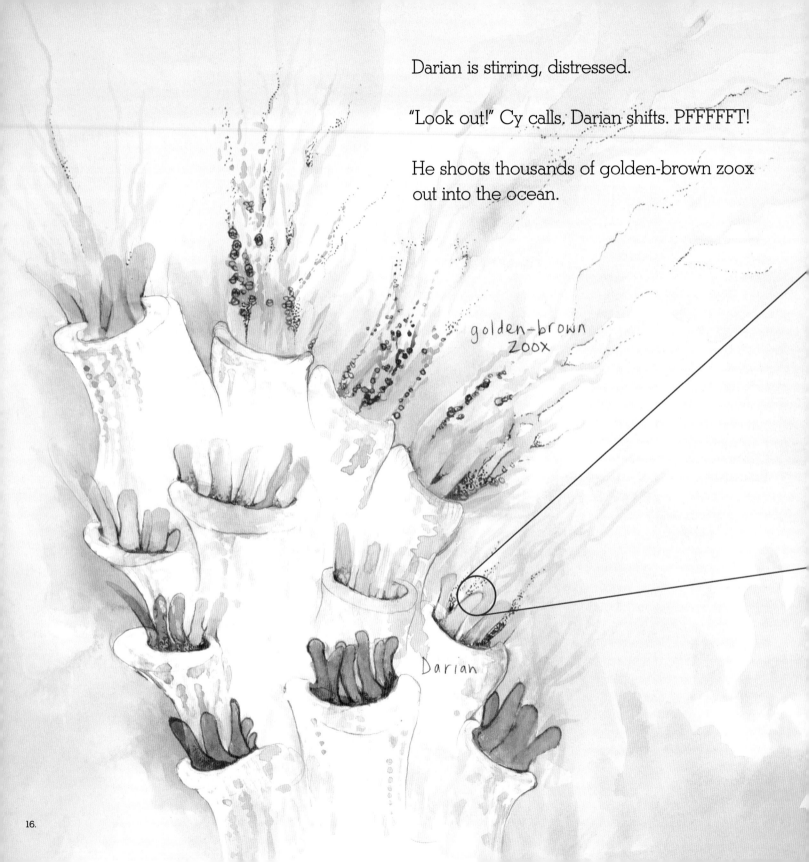

Darian is stirring, distressed.

"Look out!" Cy calls. Darian shifts. PFFFFFT!

He shoots thousands of golden-brown zoox out into the ocean.

golden-brown
zoox

Darian

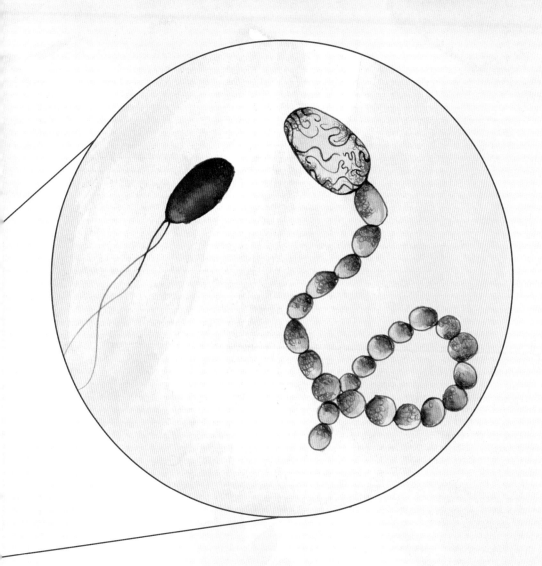

"Nooooo!" Zobi calls, as they disappear.

There are still some big zoox left, but they just keep making nasties.

Darian shoots out more and more golden-brown zoox and his tentacles begin fading to white.

"Are we going to end up like Old Pora?" Zobi asks.

"Well," Cy says, "those big zoox used to feed Darian lots of sugar, and he needs it to build our home."

"So if Darian starves…?" Zobi begins.

"He won't starve yet," Cy interrupts. "Stay up here and see."

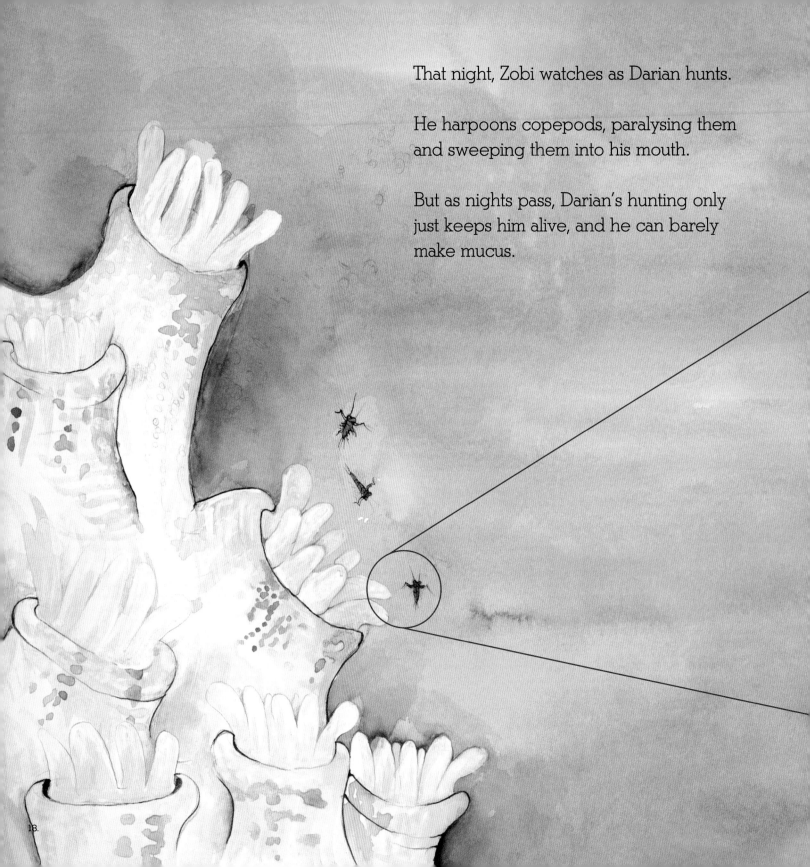

That night, Zobi watches as Darian hunts.

He harpoons copepods, paralysing them and sweeping them into his mouth.

But as nights pass, Darian's hunting only just keeps him alive, and he can barely make mucus.

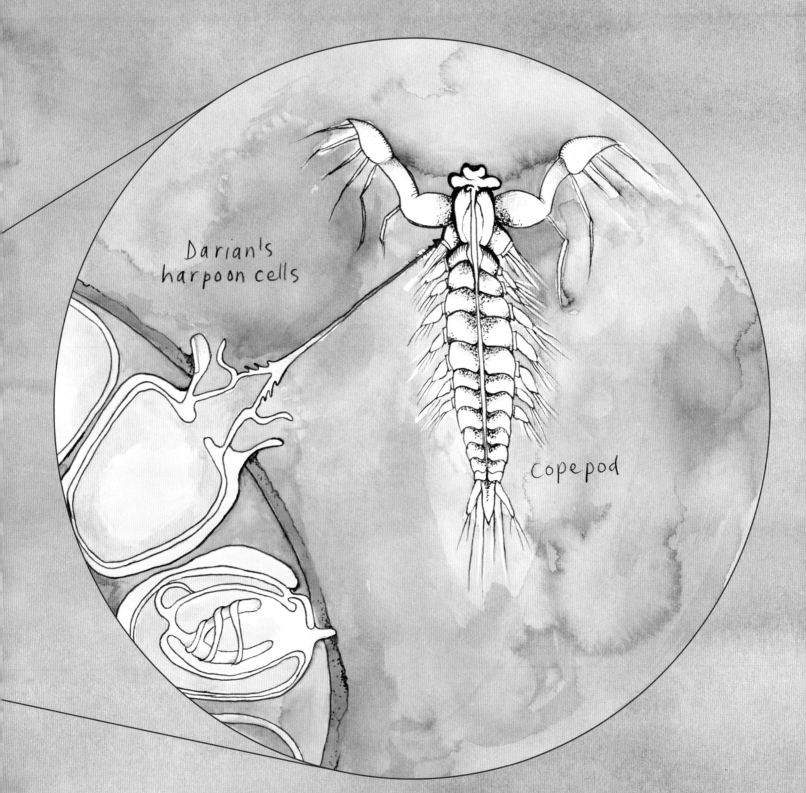

Darian's
harpoon cells

copepod

19.

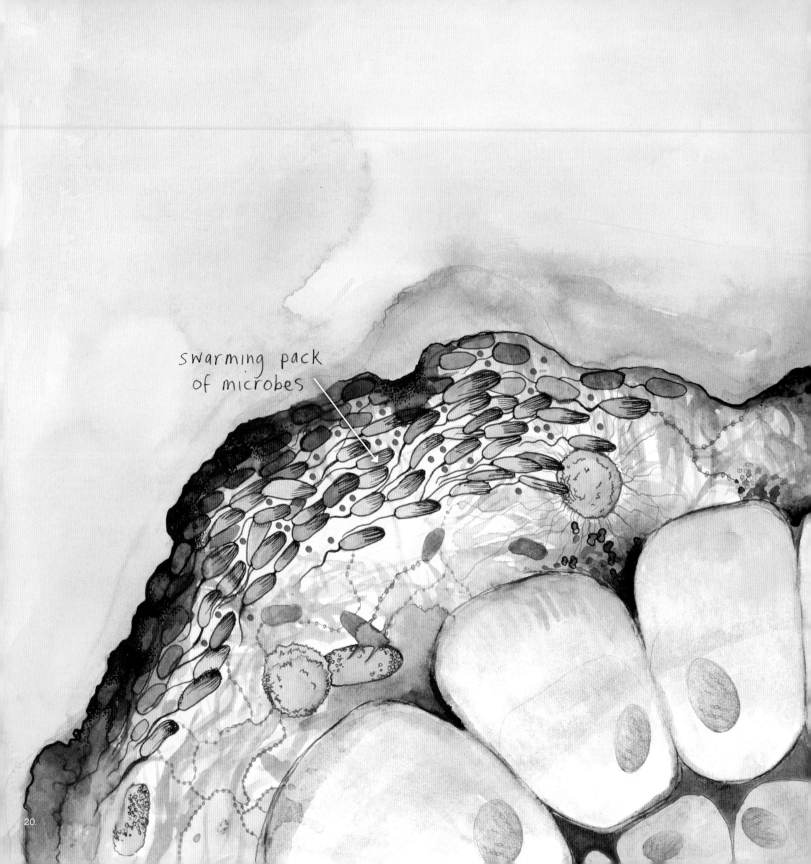

swarming pack
of microbes

The starving creatures living in the thinning mucus begin to riot. Deadly microbes start swarming in a tight pack.

The guardian bacteria can't defend the coral anymore. Zobi's family are hungry and terrified.

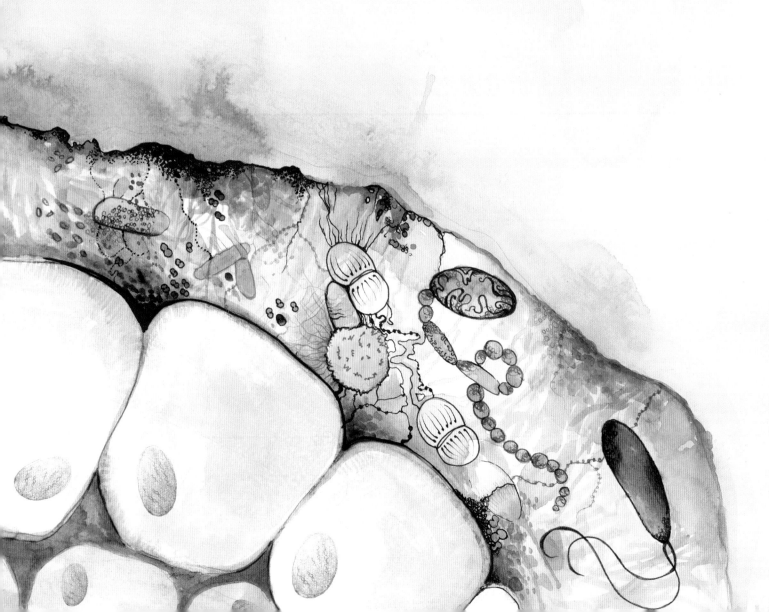

Baby algae, slimy and green, creep towards Darian, threatening to steal his sunlight.

Things are getting desperate.

slimy, green algae

Darian

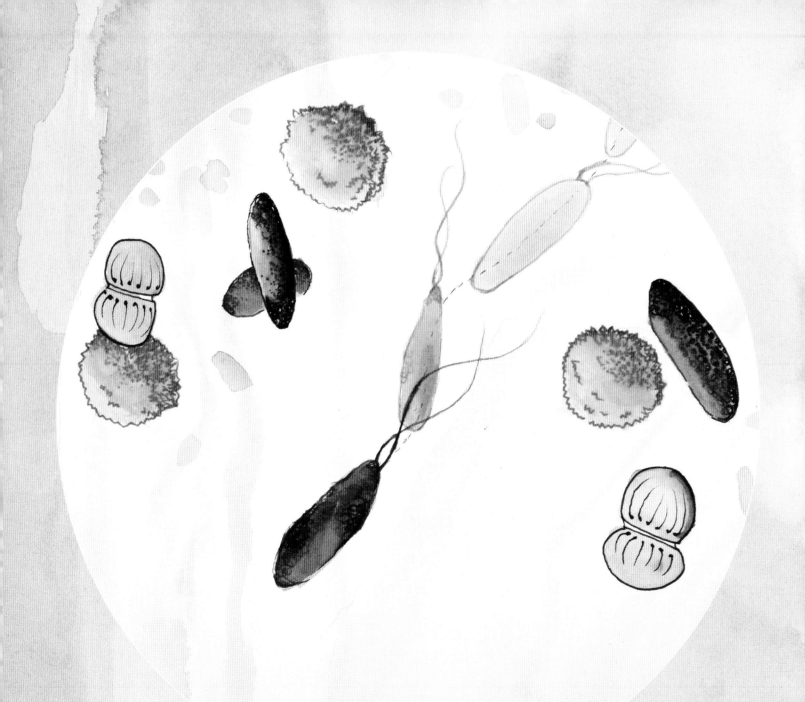

A rogue swarm of microbes rushes towards them and Zobi turns and struggles away. "Dini! Find Dini," Cy calls, "and help…"

Zobi battles down the gut, but she's weak and afraid. Around her float thousands of other starving bacteria.

When Zobi finds Dini, the little zoox
cries, "Zobi, you came!"
Zobi realises Dini is not alone.

"These are my daughters," Dini says. "We're so hungry."

They're all making sugar. But slowly, and not much.

Zobi eats a sugar molecule and finds some strength. She works her magic. Bang! Ammonia.

Dini eats it quickly and starts to look healthier.

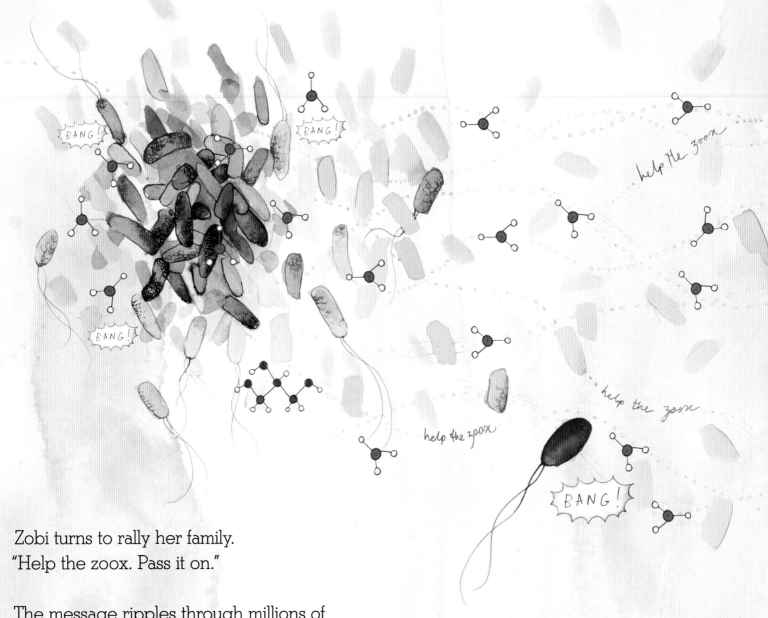

Zobi turns to rally her family.
"Help the zoox. Pass it on."

The message ripples through millions of
rhizobia. "Help the zoox. Help the zoox."

Even in the distance, on the other side
of Darian, and out through the rest of the
colony, Zobi's family struggles back to work.

Soon Dini's daughters make other daughters who begin to work. Boom! Sugar.

Zobi's family are making ammonia quickly now, and feeding it to the little zoox.

Slowly, Zobi feels the molecules swinging back into balance as Darian soaks up sugar.

The seawater is finally cooling down. Their home feels safe again. For now.

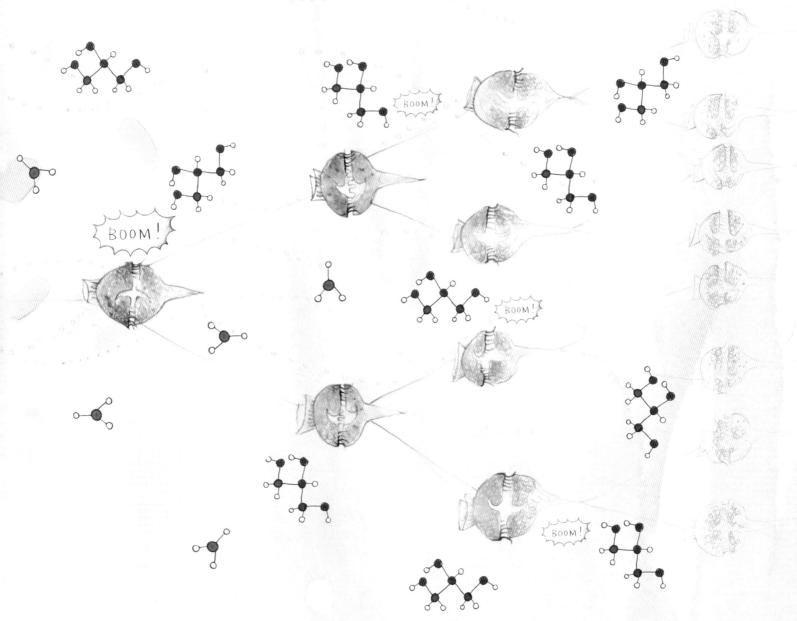

Many corals aren't so lucky.

But one by one, the survivors return to building the reef, setting it down, layer by layer, reaching for the light.

One tiny colony…

Darian and his colony

bleached coral

circling a sun.

on a small blue planet,

...in a huge blue ocean,

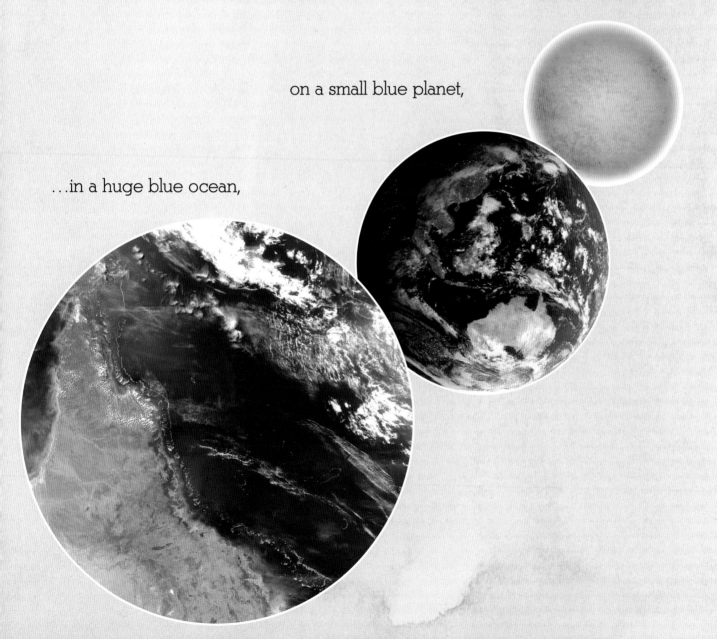

THE SCIENCE BEHIND THE STORY

A SIMPLE GUIDE TO THE CORAL SYMBIOSIS

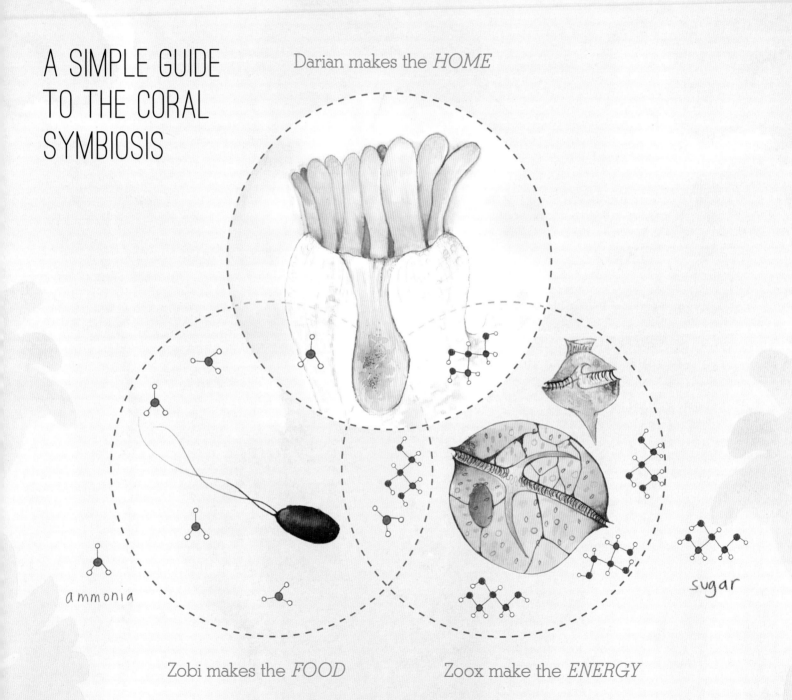

Darian makes the *HOME*

ammonia

Zobi makes the *FOOD*

Zoox make the *ENERGY*

sugar

A MAP OF DARIAN, THE CORAL POLYP

Where do the main characters live?
Darian's gut ❶, the cells inside his tentacles ❷, and his mucus ❸.

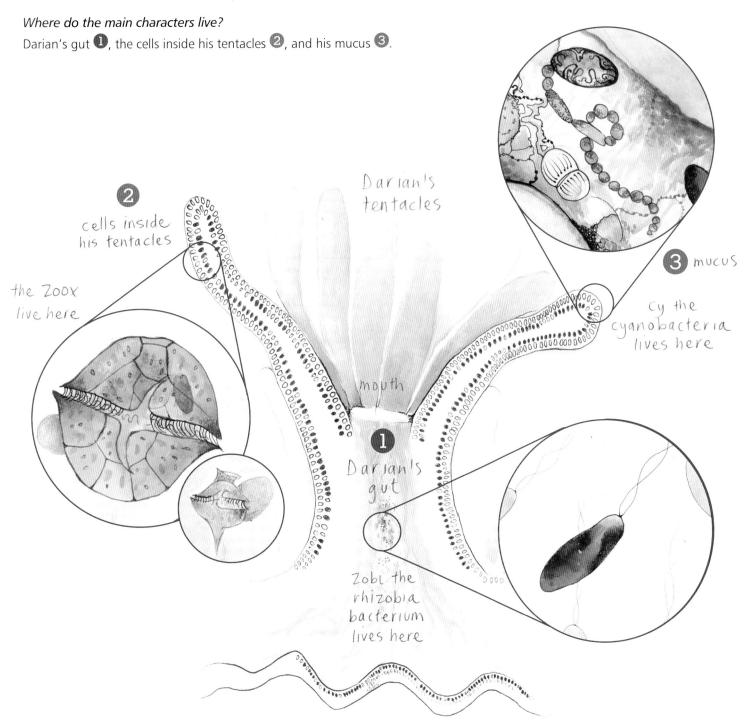

HOW SMALL ARE THE CHARACTERS?

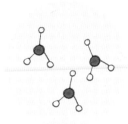

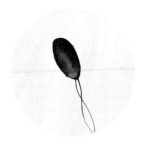

Ammonia molecule
(175 picometres)

- A molecule made from three hydrogen atoms joined to a nitrogen atom
- Essential for making proteins in all life forms
- Made by Zobi and her rhizobia bacteria family

ZOBI Rhizobia bacterium
(5 micrometres)

- Single-celled bacterium, with two flagella (tails) to help her swim
- Can make ammonia from nitrogen
- Has lived with Darian all her life

CY Cyanobacteria
(30 micrometres)

- Chain of bacteria
- Can make sugar and ammonia using sunlight and air
- One of the oldest kinds of microbes on Earth

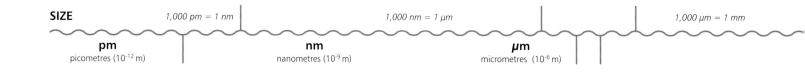

SIZE			
	1,000 pm = 1 nm	1,000 nm = 1 μm	1,000 μm = 1 mm
	pm picometres (10^{-12} m)	**nm** nanometres (10^{-9} m)	**μm** micrometres (10^{-6} m)

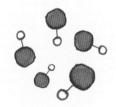

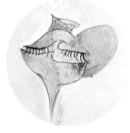

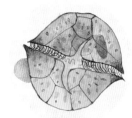

Hydroxyl molecule
(97 picometres)

- A molecule made from a hydrogen atom joined to an oxygen atom
- A common type of free radical molecule (highly reactive)
- Can damage nearly all types of molecules, including nucleic acids (e.g. DNA), proteins, fats (lipids) and carbohydrates

DINI Zooxanthellae
(15 micrometres)

- A type of single-celled dinoflagellate (algae)
- Lives symbiotically in coral polyps
- Slowly makes sugar through photosynthesis

THE ZOOX Zooxanthellae
(20 micrometres)

- A type of single-celled dinoflagellate (algae)
- Lives symbiotically in coral polyps
- Larger golden-brown type of zoox
- Makes sugar through photosynthesis

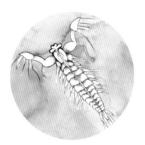

Copepods
Crustacea
(1 millimetre)

- Small crustacean (think tiny lobster), with hard exoskeleton
- Favourite food of coral polyps

Parrotfish
Scaridae
(400 millimetres)

- Large herbivorous (plant-eating) fish that eats algae from around corals
- Has a beak-like mouth

The Sun
(1,392,684 kilometres)

- Provides light energy (photons) to power photosynthesis
- 5 billion years old

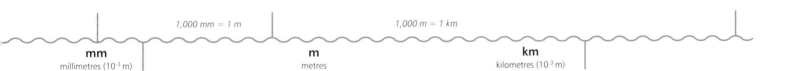

1,000 mm = 1 m   1,000 m = 1 km

mm   **m**   **km**
millimetres (10^{-3} m)   metres   kilometres (10^{3} m)

DARIAN
Coral polyp
(6 millimetres)

- The animal host in the story
- Type of stony coral (Scleractinia)
- Grows in a colony of identical polyps
- Like an upside-down jellyfish

The Doldrums
(100 kilometres)

- A weather condition defined by a lack of wind, creating still, warm patches of ocean
- Increasingly responsible for triggering coral bleaching due to climate change

ALL CORALS GREAT AND SMALL

This story is set on one of the 2,000 separate reefs which make up the Great Barrier Reef in Queensland, Australia.

Large coral reefs are found in tropical waters in many parts of the world. Their hard limestone structure is built by different species of stony corals, taking many familiar shapes including staghorn, plate, boulder, brain and mushroom.

Hard coral reefs often form large mounds made up of dozens of coral colonies, which collectively grow to the size of a car. These are often referred to as *bommies*, from the aboriginal word *bombora*.

"Darian is building the reef, layer by layer, like his family has done for hundreds of years." (pg4)

If you take a core sample from a large coral bommie, you can see a timeline of hundreds of years of growth – a bit like rings within a tree trunk. A bommie is usually a combination of many different coral species slowly growing upwards, at a rate of about 1–25 cm a year.

While some corals can grow at depths of up to 150 metres, most corals prefer to be within the top few metres of the ocean surface to allow their zoox to harvest light from the Sun. Close to the surface, the coral has to battle against the constant eroding action of waves, by setting down new layers of limestone each day.

"...lives a colony of coral polyps." (pg3)

While some polyps live as individuals, most coral polyps grow together as colonies consisting of thousands of polyps. Starting from a single, original polyp, each new identical daughter polyp cements its limestone skeleton to its neighbour, connected to each other through tubes.

reef

bommie

colony

polyp

DARIAN IN DETAIL

A coral is like an upside-down jellyfish.

Along with jellyfish, anemones and hydroids, corals belong within the animal kingdom in the Phylum *Cnidaria*. The word *Cnidaria* (ny-DAIR-ree-ah) comes from the Greek *knide*, meaning stinging nettle. All cnidarians have only two cell layers, radial symmetry, simple muscle and nervous systems and a single mouth-like opening for eating.

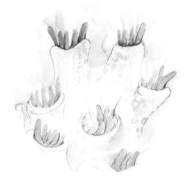

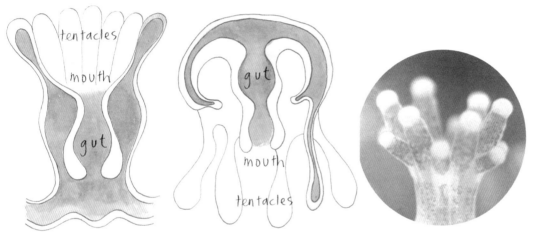

Coral polyp

jellyfish

zoox in a polyp

"He harpoons copepods, paralysing them and sweeping them into his mouth." (pg 18)

The stinging cells of a coral are called a cnidocyte or nematocyte. When triggered by movement of prey in the water, they fire a harpoon-like structure containing a toxin to paralyse prey, such as a tiny copepod, so they can reel it in and swallow it.

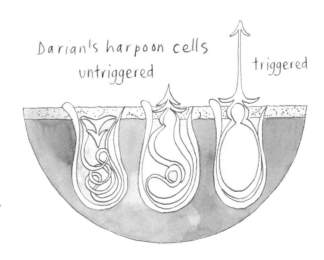

Darian's harpoon cells untriggered | triggered

How does Darian build the reef?

"...those big zoox used to feed Darian lots of sugar, and he needs it to build our home." (pg 17)

Many corals rely on a steady supply of sugar from their photosynthetic zooxanthellae (zoox) partners to give them the energy for their basic needs.

Different coral cells use this energy to power a range of tasks, such as making mucus, harpooning prey or building their limestone home.

Inside a healthy, well-lit coral, zoox are so good at making sugar, there is usually plenty to spare building lots of hard limestone reef (calcium carbonate).

Where does Darian fit in the scientific classification of all animals?

Domain: *Eukarya*
Kingdom: *Animalia*
Phylum: *Cnidaria*
Class: *Anthozoa*
Order: *Scleractinia*
Family: *Acroporidae*
Genus: *Acropora*
Species: *tenuis*

ENERGY FROM THE SUN

Almost all life on Earth relies on energy from the Sun. This energy is captured by plants, algae (including zoox) and some types of bacteria through the process of photosynthesis.

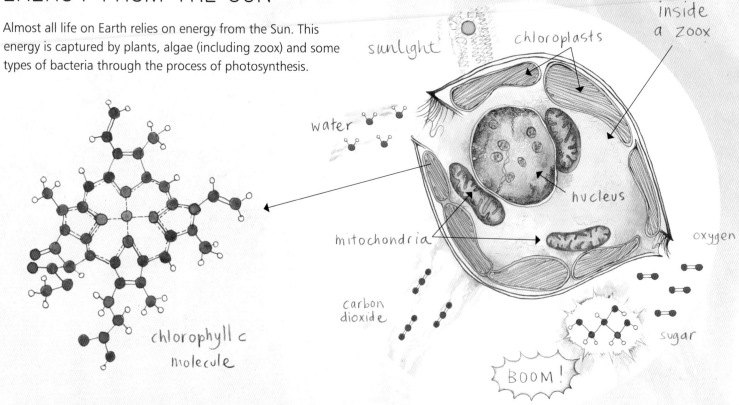

chlorophyll c
molecule

sunlight

water

carbon
dioxide

BOOM!

inside
a zoox

chloroplasts

nucleus

mitochondria

oxygen

sugar

PHOTOSYNTHESIS

"They pull down sunlight, stir in water, and add carbon dioxide. Boom! Sugar. Oxygen bubbles around them." (pg12)

The process of photosynthesis relies on a steady stream of light (photons) from the Sun. This light excites electrons in the chlorophyll molecules inside the chloroplasts. These electrons are used to split water (H_2O) into separate hydrogens (H) and oxygens (O).

Through photosynthesis, many of the oxygen atoms from water are released as di-oxygen (O_2) bubbles. The hydrogens are used to power the final major step – the transformation of carbon dioxide molecules into different types of sugar. The sugar given to the coral from the zoox is a three-carbon sugar called glycerol.

Put simply, the photosynthesis reaction is:
light + water + carbon dioxide = oxygen + sugar

light + water + carbon dioxide ⇒ oxygen + sugar (glycerol)

THE ZOOX

The name zooxanthellae comes from the Latin *zoo* (of animal) and *xantho* (yellow) plus *ella* (little). Commonly referred to as "zoox", these single-celled microscopic creatures can photosynthesise using slightly different mixtures of pigment molecules. This gives them characteristic shades of golden-brown, rather than the typical green colour of plants.

Even though zoox can photosynthesise like plants and are named with the prefix "zoo", they are neither plants nor animals. Instead, they are currently thought to belong to the Kingdom *Chromalveolata*, one of several Kingdoms (including *Plantae, Animalia, Fungi*) that together make up the Domain *Eukarya* (life forms whose cells have a nucleus). Zoox share their kingdom with many weird and wonderful tiny creatures, such as amoebae and slime moulds.

All zoox are dinoflagellates (a type of algae), belonging to the single genus *Symbiodinium* (symbiotic dinoflagellates). These single-celled microscopic creatures are "eaten" by the coral, but instead of digesting them, the coral welcomes them to reside inside some of its cells. The coral provides them with a safe home and (usually) the right amount of sunlight, and they, in turn, make sugar for the coral. After taking up residence in a coral cell, these zoox lose their flagella (tail).

Domain: *Eukarya*
Kingdom: *Chromalveolata*
Phylum: *Dinoflagellata*
Class: *Dinophyceae*
Order: *Suessiales*
Family: *Symbiodiniaceae*
Genus: *Symbiodinium*
Species: *unknown*

"The nasty molecules are hurting Darian... they're coming from the big zoox." (pg15)

A natural by-product of photosynthesis by the zoox is the creation of free radicals – molecules such as superoxide (O_2^-) and hydroxyl (OH). Free radicals are powerfully toxic because they can easily damage other molecules, including proteins and DNA. Most cells (including coral) have the ability to deactivate small amounts of free radicals. However, in prolonged periods with high sunlight, large amounts of free radicals build up – causing damage to DNA, leading to irreversible mutations and potential cell death. This is why coral cells will try to protect themselves under conditions of high stress and often expel their symbiotic zoox.

"Zobi notices one zoox nearby, smaller than the rest." (pg13)

There are different kinds (clades) of zoox within the genus *Symbiodinium*. Some zoox are different sizes and some have different photosynthetic pigments, meaning that some perform photosynthesis much more efficiently (faster) than others. And some (like Dini in our story), perform photosynthesis more slowly.

zoox under a microscope

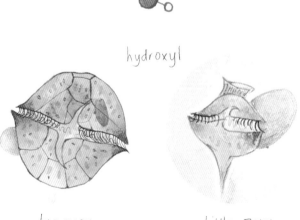

hydroxyl

big zoox

little zoox

BRINGING NITROGEN TO LIFE

The element nitrogen is an essential building block for all life on Earth. It is used to make amino acids in proteins and building blocks for DNA and RNA. The atmosphere and seawater contain large amounts of nitrogen (as di-nitrogen, N_2), but because this molecule is tightly held together by strong triple bonds, it is impossible to use.

Luckily, microorganisms like rhizobia (e.g. Zobi) and *Cyanobacteria* are experts in breaking these nitrogen bonds – and are therefore vital to all life on Earth. They do this by converting the di-nitrogen (N_2) into ammonia (NH_3). This process is called nitrogen fixation.

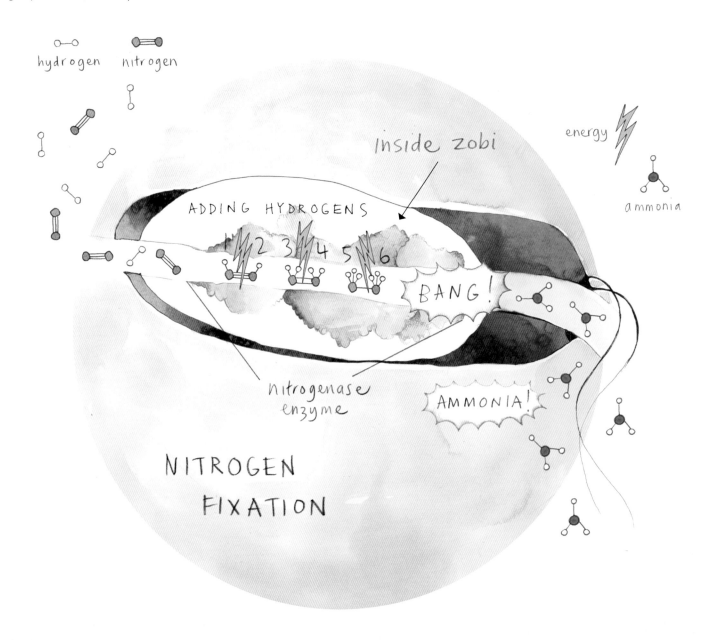

hydrogen   nitrogen

inside zobi

energy

ADDING HYDROGENS

1 2 3 4 5 6

BANG!

ammonia

nitrogenase
enzyme

AMMONIA!

NITROGEN

FIXATION

"She takes a pair of nitrogens and adds hydrogens." (pg7)

The process of nitrogen fixation converts atmospheric di-nitrogen (N_2) into ammonia (NH_3) through the equation:

$$N_2 + 8H + energy = 2NH_3 + H_2$$

Only a handful of different microbes have ever gained the nitrogenase enzyme, which allows them to fix nitrogen. This select group includes some Archaea and a few different groups of Bacteria, including *Vibrio*, *Cyanobacteria* and rhizobia.

"Ammonia. Dini eats it quickly and starts to look healthier." (pg25)

All life forms need a regular source of nitrogen, such as ammonia. Coral polyps can obtain some nitrogen by hunting and digesting small plankton (such as copepods and algae), breaking down their proteins into simple nitrogen-based units such as amino acids. But scavenging nitrogen like this can only go so far when coral polyps have many small mouths to feed within their symbiotic community – especially their hungry zooxanthellae (like Dini). Some coral polyps can also host rhizobia bacteria (like Zobi), who can fix nitrogen (in the form of ammonia) for the coral in return for food and a safe home. This symbiotic relationship is much like those of rhizobia bacteria inside the roots of legume plants in the soil.

ZOBI THE RHIZOBIA

"Zobi and her bacterial family are making food for Darian." (pg7)

Within the first few days of their life, many corals already have rhizobia bacteria inside them, fixing nitrogen. What's more, when needed, these rhizobia can even make extra ammonia for their coral host. Perhaps this helps explain why corals have survived for over 240 million years.

Domain: *Bacteria*
Phylum: *Proteobacteria*
Class: *Alphaproteobacteria*
Order: *Rhizobiales*
Family: *Bradyrhizobiaceae*
Genus: *Bradyrhizobium*
Species: *japonicum*

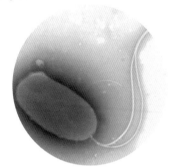

CY THE CYANOBACTERIA

"Cy is wise and old." (pg10)

Cyanobacteria are more ancient (about 3 billion years old) than coral (about 240 million years old). Many *Cyanobacteria* grow in chains (filaments), dozens to hundreds of cells long. Some cells can fix both nitrogen (like rhizobia) and carbon through photosynthesis (like the zoox). This amazing combination might explain why some types of *Cyanobacteria* have even been found inside coral cells, replacing the zoox!

Domain: *Bacteria*
Phylum: *Cyanobacteria*
Class: *Cyanophyceae*
Order: *Nostocales*
Family: *Nostocaceae*
Genus: *Nodularia*
Species: *spumigena*

MUCUS IS A MATRIX OF MICROBES AND MOLECULES

Coral, like all animals, makes mucus. We humans make lots of it every day in our bodies, such as our nose and our gut. Mucus is a matrix of molecules built from branching sugars attached to a protein backbone. This sticky substance is constantly secreted by special cells in the coral (mucocytes) and helps protect the coral polyp from random attacks by microbes in the seawater like a shield.

"Trillions of small friends live in and on Darian." (pg5)

A single coral polyp plays host to trillions of microorganisms (microbes), both inside its gut cavity and in its rich layer of mucus surrounding its mouth and tentacles. The microscopic residents within this coral society include viruses, bacteria, archaea, fungi, protists and a variety of algae (including dinoflagellates), many of whom create sugars, proteins and DNA molecules, further adding to the sticky mucus matrix.

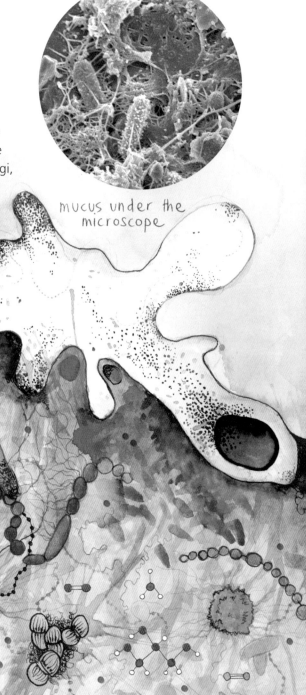

mucus under the microscope

TDA
(weapon molecules)

40

"Zobi dodges between microbes who are eating, recycling and swapping molecules." (pg9)

Coral reefs around the world thrive in shallow seawater environments where there are very little nutrients available. Within each coral polyp, microbes play a crucial role in recycling the molecules of all nutrients containing nitrogen, sulfur and phosphorus. Nothing is wasted (it's a little like living in a space station).

One of these sulfur-containing molecules – dimethylsulfide (DMS), released by some coral bacteria – can travel into the atmosphere (as an aerosol), where it is thought to seed the formation of clouds.

"She passes a crew of guardian bacteria making deadly weapons to defend Darian." (pg9)

Some of the sulfur-containing molecules are recycled by some coral bacteria (called *Pseudovibrio*) into potent antimicrobials called TropoDithietic Acid (TDA), which is thought to help defend the coral from outside invaders.

"The starving creatures living in the thinning mucus begin to riot." (pg21)

When the microbial community in the coral mucus falls out of balance (such as when there is too little mucus to share), bacteria such as *Serratia marcescens* can coordinate and swarm using Acyl-Homoserine Lactone (AHL) molecules, causing coral diseases such as white pox.

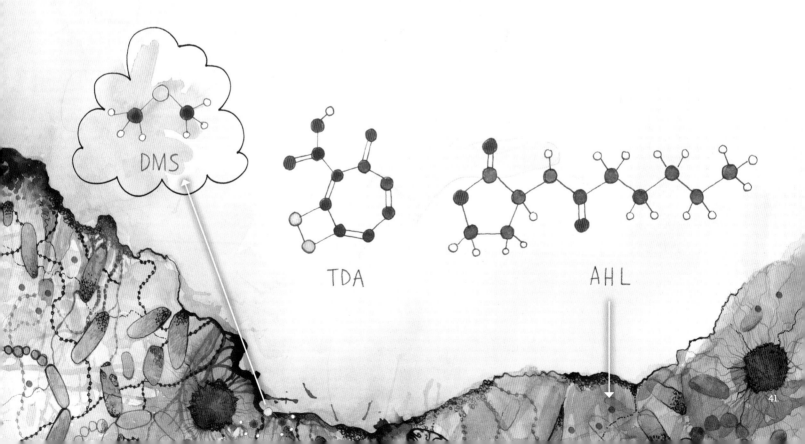

DMS

TDA

AHL

CORAL BLEACHING

Coral bleaching is the massive loss of the photosynthesising zooxanthellae from most coral polyps within a coral colony. It's triggered by environmental stresses including physical extremes of temperature (both too hot and too cold), extreme UV light, chemical stress caused by fertiliser or herbicide runoff from farms, sediment from land clearing, freshwater runoff after cyclones and some sunscreen chemicals.

In this story, the coral bleaching is caused by heat stress. If levels of heat stress are not too extreme (<2°C above summer average) and short lived (a month or less), then most bleached corals will recover. If conditions are more extreme (>3°C) for more than 6 weeks, most corals will not survive.

+2°C triggers bleaching

TIMELINE

The timeline for the story *Zobi and the Zoox* is set across several weeks.

DAY 1

"The ocean is still and it's far too hot." (pg5)

Bleaching is usually caused by 2–3 weeks of low wind conditions (the Doldrums), combined with hot summer sunlight, causing seawater temperatures to rise by 1–4°C. Tragically, this extreme weather pattern appears to be increasingly occurring in late summer months on the Great Barrier Reef and other reefs around the world.

DAY 10

"The nasty molecules are hurting Darian... they're coming from the big zoox." (pg15)

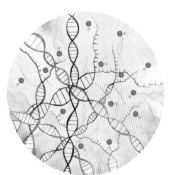

Why corals destroy and/or eject their zoox is still the subject of research. It is thought that under prolonged periods of stress, the zoox slowly cause a build-up of toxic stress molecules (free radicals) that can cause damage to vital proteins and DNA inside the coral cells.

DAY 15

"He shoots thousands of golden-brown zoox out into the ocean." (pg16)

The coral can normally mop up small amounts of these toxic molecules with other molecules called antioxidants. But when it is overwhelmed by too many free radicals, the coral is programmed to start removing the zooxanthellae from inside its cells.

DAY 20

"...and his tentacles begin fading to white." (pg17)

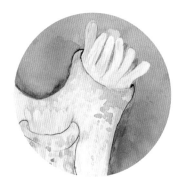

Once the coral polyps have ejected most of their colourful zooxanthellae, their clear tissue will show through to the hard limestone underneath, making them appear white.

CAN DIVERSITY SAVE THE REEF?

Not all zoox are the same. The zooxanthellae genus *Symbiodinium* has nine broad genetic groups (A–I) called clades. The dominant type of zoox on the Great Barrier Reef is clade C (the big zoox in our story). Scientists have discovered that corals containing clade D zoox (Dini and her daughters in our story) are more tolerant of higher temperatures. This is possibly because the slower rate of clade D photosynthesis places less stress on the coral during hot bleaching conditions. While the diverse abilities of different zoox offer coral reefs some hope for adapting against our changing climate, this amazing adaptability might only save some corals from small amounts of heat stress. We still need to work quickly to remove other human stresses on coral.

DAY 25

"...those big zoox used to feed Darian lots of sugar, and he needs it to build our home." (pg17)

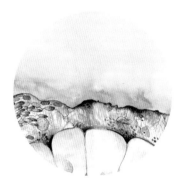

The zoox were the coral's main source of energy – needed to grow protective mucus and to continue to build the hard reef. In order to survive, each coral polyp must nurture the recovery of the remaining zoox like Dini... and possibly even recruit new zoox from the surrounding seawater.

SEVERAL WEEKS/MONTHS

"Baby algae, slimy and green, creep towards Darian, threatening to steal his sunlight." (pg22)

Algae and seaweeds surrounding coral grow more quickly in warmer seawater. If the seawater temperature remains high after coral bleaching, algal growth will place even more pressure on the coral community, making it even more difficult to recover.

FUTURE RECOVERY   or

"Soon Dini's daughters make other daughters who begin to work. Boom! Sugar." (pg27)

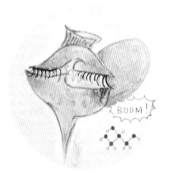

Towards the story's end there is a quick recovery of Dini's family of little zoox. However, a full recovery from bleaching needs lots of energy and can take up to a year.

FUTURE DEATH

"A broken pile, covered with thick, dirty algae." (pg11)

If the coral colony does not quickly recover a healthy population of zoox, it will likely die of starvation, and ultimately become overgrown by slimy algae and seaweeds (like Old Pora).

GLOSSARY

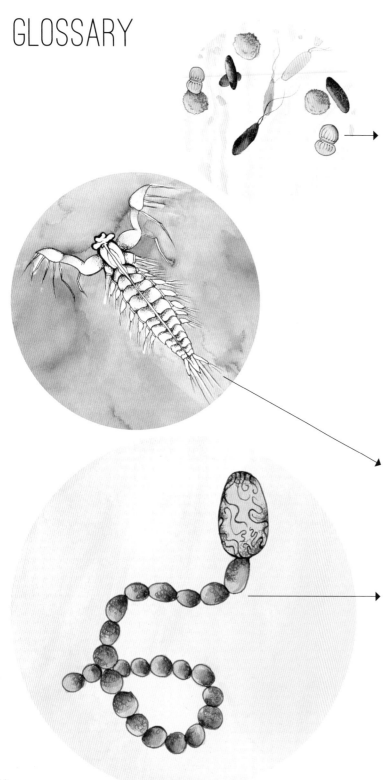

AMOEBA

A type of single-celled organism found in all types of water – from soils and puddles to rivers and the open ocean. An amoeba (plural amoebae) can move and feed using extremely flexible arms of their membrane, called pseudopods.

BACTERIA

The smallest single-celled life form. They are usually about 1 or 2 micrometres long (there are 1,000 micrometres in a millimetre). Scientists have classified thousands of different species of bacteria (singular bacterium), but it is thought there could be millions – we just don't know about them yet. They come in many different shapes and sizes, including bacillus (rod-shaped), coccus (spherical), spirillum (spiral-shaped) and vibrioid (comma-shaped).

CELL

The word cell comes from the Latin *cella*, which means small room. A cell is the basic self-reproducing building block of life. Some organisms such as bacteria and dinoflagellates (e.g. zooxanthellae) are only made of one single cell (uni-cellular), while larger organisms such as coral, fish, seaweed and humans are made of many cells (multi-cellular).

COPEPOD

Tiny crustaceans found in both saltwater and freshwater. There are at least 13,000 species of copepods, all of whom feed on small plankton. They have a hard exoskeleton like all crustaceans (such as shrimp and crabs), but because they are so small, they are usually transparent.

CYANOBACTERIA

The largest group of photosynthetic bacteria. Also, one of the most ancient groups of bacteria – scientists have found fossils of *Cyanobacteria* (or blue-green algae) that are more than 3 billion years old. Like Cy, many *Cyanobacteria* form chains, which make them look somewhat like a worm, but this filamentous structure is actually many cells joined together. Joined like this, the microbes are often mobile and can be found gliding through the ocean. A billion years ago, *Cyanobacteria* became the photosynthetic organelles (chloroplasts) inside the cells of all modern plants.

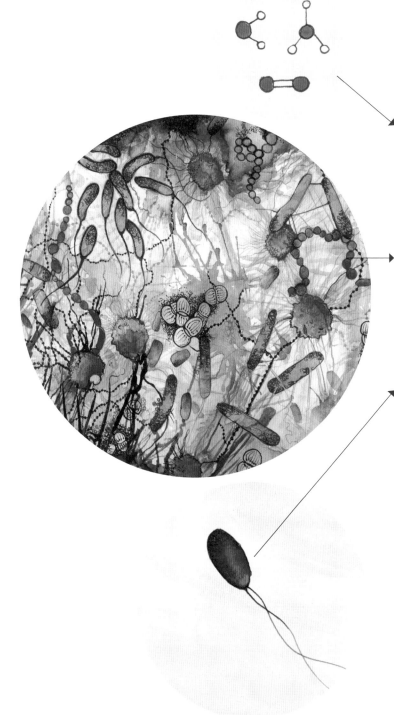

DNA

DeoxyriboNucleic Acid (DNA) is the macro-molecule used to encode genetic instructions in all cellular life forms.

MICROBE

Short for 'microorganism', it refers to microscopic life forms, such as bacteria, archaea, viruses and protozoa.

MOLECULE

A molecule is a group of two or more atoms held together by chemical bonds. Some molecules, such as oxygen (O_2), water (H_2O) and ammonia (NH_3) are simple and small. However, even a structure as complex as DNA can also be described as a type of molecule.

MUCUS

Healthy corals cover all of their external surfaces with a thick layer of mucus – a watery slime, containing sugars and proteins. All animals (including corals) make mucus to nurture good microbes and protect against pathogens (infectious microbes) and particles (like dust).

RHIZOBIA

The group of nitrogen-fixing bacteria that can symbiotically partner with the roots of legume plants in soil and some corals in the ocean. They are essential for converting (fixing) atmospheric di-nitrogen (N_2) into usable forms of nitrogen, such as ammonia (NH_3).

ZOOX

Small single-celled dinoflagellates from the genus *Symbiodinium*. Zoox (short for zooxanthellae) are one of the most successful organisms on Earth, able to symbiotically partner with a range of marine organisms, from corals, anemones and jellyfish to giant clams, sponges and tiny protozoa.

SYMBIOSIS ON THE REEF

Coral reefs are amongst the most diverse ecosystems on Earth. Occupying less than 0.1% of the world's ocean surface, they are home to over 25% of all marine species.

Every square centimetre of coral reef is populated by living organisms. The hard coral structure provides the perfect physical environment for all of the diversity found on the reef. This includes: soft corals, sponges, fish, anemones, worms, shrimp, sharks, clams, squid, starfish, urchins, sea cucumbers, tunicates, turtles and sea snakes.

Coral reefs thrive in very low nutrient environments, where there is very little organic matter to feed on. When Charles Darwin first described coral reefs in 1842, he was confused about how they could exist in low nutrient conditions. For over 100 years, this was called Darwin's Paradox.

How do so many life forms exist in such challenging conditions?
By working together.

Beneficial symbiotic partnerships like the one in our story are common in coral reefs. The low nutrient conditions make it essential for life forms to work together efficiently, to use what little food is available. Some of the many examples of symbiosis include: clownfish with anemones; cleaner shrimp with fish; sea slugs solar-powered by Cyanobacteria; and sponges – the microbial hotels of the reef.

But at the heart of all coral reefs – and this story – is the partnership between coral polyps, zooxanthellae and bacteria. The ability of the zoox and rhizobia bacteria to create sugar and ammonia creates vital nutrients at the base of the reef food web.